GARDEN FLOWER SEEDS
A Pictorial Field Guide

I0413952

Terry A. Woodger

Universal-Publishers
Boca Raton

Garden Flower Seeds:
A Pictorial Field Guide

Universal-Publishers
Boca Raton, Florida
USA • 2011

ISBN-10: 1-61233-041-X
ISBN-13: 978-1-61233-041-9

www.universal-publishers.com

Library of Congress Cataloging-in-Publication Data

Woodger, Terry A.
  Garden flower seeds : a pictorial field guide / Terry A. Woodger.
    p. cm.
  Includes bibliographical references and index.
  ISBN-13: 978-1-61233-041-9 (pbk. : alk. paper)
  ISBN-10: 1-61233-041-X (pbk. : alk. paper)
  1. Flowers--Seeds--Harvesting. 2. Seeds--Cleaning. 3. Flowers--Seeds--
Storage. 4. Flowers--Seeds--Pictorial works. 5. Flower gardening. I.
Title.
  SB118.3.W66 2011
  631.5'21--dc23

                                                    2011032344

ACKNOWLEDGMENTS

A book of this nature can not be written without the assistance of family and friends. I would like to acknowledge the following people who assisted in many ways to help make this book a reality.

First, I'd like to thank my wife and children for their support and encouragement, without which this book would never have been completed.

I'd also like to thank the following people for their assistance:

Andrew Leighton
Clive and Cathy Grimshaw
Colin and Shirly Cattle
Eddy and Marion Pettifor
Graham and Kathy Evans
Jacqueline Weight
Ron and Lynda Roundhill
Tim and David Evans

Disclaimer:
Plants have many ways in which they protect themselves from damaging organisms. This protection is found in thorns, sap, toxins, etc. Although the collection, cleaning, and storage of seed can be a rewarding experience, the author takes no responsibility for injury or illness that results from these activities.

CONTENTS

INTRODUCTION

Seeds are an exciting and beautiful component of flower gardening. The vast gardens that can be created from just a handful of these treasures can give a great sense of achievement to a gardener.

Although some of these plants can propagate by other means, such as bulbs or division, seeds are the principal way in which garden flowers reproduce.

This book covers the basics involved in the collection, cleaning and storage of seeds. Although bulbs, corms, and other plant parts can be collected and stored, they are not covered here, so as to not detract from the focus of this volume.

As the number of plants grown in flower gardens is truly staggering, it is impossible to cover them all. In this field guide, we discuss the most common plant families, including examples of the types of seeds that may be encountered. Where possible, several genera within each family are discussed.

This book uses a system whereby plant family names are all written in capitals (AGAVACEAE), the common names that are not written within the text are in bold (**Agave**), and the

botanical names are written in italics (*Agave attenuata*).

In botany, it is the characteristics of the flowers that determine the genera and family to which a plant belongs. This can become extremely complicated, so this field guide makes no mention of the flower types or their individual differences.

Also discussed are a number of methods that can be utilized in the collection of seeds. No one method can be used to collect them all, so different techniques have been developed over time to successfully gather all of the species that are encountered, both in the home garden as well as in the field.

The same development of techniques applies to the cleaning of seeds. There are a number of ways in which common household items can be used effectively to clean seeds. Several of these items are explained in Chapter 3: How to Thresh and Clean Seed.

Storing seeds for use next season can be fraught with hidden problems, such as molds and seed-borers. Chapter 4: The Storage of Seed explains some appropriate methods and procedures that should be followed to avoid disappointment and loss of seed.

Most seeds collected from the garden are suitable for storage from one year to the next, and many of these can be successfully stored at home for many years.

Some elements to consider when collecting seeds are the quantity and the number of plants from which the seeds are collected.

Collecting seed from only one fruit on one plant over several seasons can have unforeseen consequences, such as only ending up with seeds from small fruit. This can produce smaller plants with smaller fruit.

This is often referred to as "line-breeding" and can, over time, lead to complete genetic breakdown and a loss of viability.

The collection of seed is not always a simple matter, and it is very important that, when possible, you gather seed from several plants to maintain their genetic stability.

Size does matter, and in the case of growing flowers, it is always best to collect seed from the biggest fruit or healthiest plants.

The exception would be if you are after a specific genetic trait. For example, seed selection would vary for a plant desired as a bush if that plant is usually a climbing variety.

Now let's look at the question of, what is a fruit?

When considering plants, a fruit is any structure that produces seed, but not always something that is edible. The following pages outline the types of fruit that will be encountered in the collection of seeds from herbs and spices.

One of the first things that should be learned in the collection of seed is the differences between the various fruits that you will encounter, as this will determine how to collect and clean the seed.

The methods used for the collection of seeds do not require an intimate knowledge of fruits. Knowing the basics as outlined here will aid in the collection, cleaning, and storage of viable seed.

Achene: An achene is a small, single-seeded, dry fruit usually formed in clusters of the ASTERACEAE (daisy) family. Fruit of this group include the sunflower, marigold, and the dandelion.

Marigold
Calendula officinalis

Berry: A berry is a fruit with seeds contained within a fleshy or dry pulp. Berries contain two or more seeds, and include the cactus and duranta.

Torch Thistle *Cereus* sp.

Burr: A burr is a spiny fruit that can cling to passing objects as an aid in dispersal. Burrs are found in numerous plant families and include species of amaranth, aster, and daisy.

Cobbler's Peg *Bidens* sp.

Capsule: A capsule is a dry seed case that opens along several seams and can be papery thin to hard and woody. Some open at maturity, releasing their seeds, while others remain closed. Capsules can contain anything from one to over one hundred seeds and are found in many plant families.

The *Aristolochia* contains the seeds within a capsule which resembles an upside-down parachute.

Indian Birthwort
Aristolocia tegarla

Drupe: A drupe is a fleshy or dry fruit containing one seed, often referred to as a "stone," hence the origin of the term "stone fruits." These include the cook tree, dracaena, and fireball.

Fireball
Haemanthus multiflorus

Grain: "Grain" is a term used to describe the husk-covered seeds from the family POACEAE (grasses). Many of these plants are used in flower gardens for their attractive foliage.

Hip: A hip is a term used to describe the fruit of a rose. It is essentially a dry berry.

Rose Hip *Rosa* sp.

Legume: Legumes split into two equal halves upon drying. Legumes include all the beans and peas. They can be thin and papery, or thick and woody. The dolichos bean and butterfly pea are examples of legumes.

Dolichos Bean
Dolichos purpureus

Nutlet: A nutlet is a very small nut, often found in grasses such as this sedge.

Sedge *Cyperus* sp.

Silique: These fruits are all from the mustard family (BRASSICACEAE). All have dry, woody pods with a partition down the center. The pods open at maturity and release their seeds as they begin to dry.

Mustard *Brassica juncea*

CHAPTER 1
THE COLLECTION
OF SEED

Seed should be collected when the weather is fine and the plants are dry.

If you have been able to collect seed pods or capsules when they are dry, then it is only a matter of cleaning. However, if the collected material is wet or damp, then mold may become an issue. Wet collected material can be spread out on a tarp in the sun to dry. If this is not possible, spread them near some sort of warmth, such as a heater, and allow them to fully dry prior to cleaning.

Removing seeds from fleshy berries and drupes is usually a simple matter if they are fully mature.

When transporting seed overseas or across borders, a declaration of the species and amounts is often required, as some species are prohibited. If the seed is not thoroughly cleaned it may be confiscated.

When seed is required for personal use and cleanliness is not so important, keep in mind that insects and molds can still become a problem, so it remains important to clean all seed as thoroughly as possible.

There are a number of ways to collect seed; it is a matter of selecting the one which works best for you and the plant species from which you are collecting.

The use of a bucket or two when collecting fruits, capsules and flower heads is essential. Collecting large amounts of seed by hand or only small amounts from a few plants can become cumbersome without something to put it in; you can use bags, but they can often become more annoying than helpful unless you put the bags in the bucket.

Many garden flowers produce flower stalks that make collection easy. You can gather them in two ways, depending on which one is most convenient to the requirements of the individual. You can harvest the individual seed capsules as they ripen or you can place a material bag over the entire seed head and tie the mouth of the bag off. The seed head can be cut off immediately or left for a while on the plant. The risk of leaving seed heads on the plant is moisture from rain or irrigation. If the seed becomes wet, mold and spoilage will become a major threat.

Another useful method is to place a ground sheet under the plant and tap or shake the plant so that the seeds fall onto the ground sheet. The severity of the tap or shake will often depend on the maturity of the seed and the species of plant.

Avoid collecting seed or damaged fruit from the ground as they can be contaminated with mold or insects.

By following the methods outlined for each flower variety, the collection of viable seed should be achievable throughout the seasons.

CHAPTER 2
GARDEN FLOWERS

Garden flowers represent the plants used by gardening enthusiasts around the world to beautify landscapes both small and large. The number of flowering and foliage plants used in landscaping throughout the world is considerable. Therefore, trees, shrubs, and vines are covered in separate chapters. Only the commonly-recognized plants used in landscape gardening are covered in this chapter.

The collection of seed from flowering plants involves the collection of fruits, such as small berries, drupes, capsules, pods, and nuts. Unlike edible fruits that are large, colorful, and highly visible, many of these fruits are hidden from the view of the casual observer.

Some of the plants used in landscape gardening are poisonous, so care must be taken when collecting seed. Harvest the fruits and seeds using the various methods outlined in the introduction to collecting. Collect only the largest and healthiest specimens to ensure the highest quality seed is saved.

Family: ACANTHACEAE
Common name: **Acanthus**
Number of genera: 250
Number of species: 2500
Origin: Tropical
Plants: Herbs and shrubs with some climbers

The fruit is a capsule that often ejects its seed explosively as it dries lengthwise along the capsule. The capsules contain two to numerous seeds that are minute in size to large.

The flattened oval seed can be hairy, smooth, furrowed or sculptured. The seeds are yellow, grey, brown, or black.

Collect the capsules as they begin to change color, being careful to place them in a bag or container, as the seed of some species can be expelled some distance from the capsule when it opens violently.

Forest Bellbush *Asystasia bella*	**Chinese Violet** *Asystasia gangetica*
Shrimp Plant *Justicia brandegeana*	**Clock Vine** *Thunbergia arnhemica*

Family: AGAVACEAE
Common name: **Agave**
Number of genera: 18
Number of species: 600
Origin: Tropical and subtropical
Plants: Evergreen perennials

The fruit is a capsule or berry; the seed is a flattened oval shape to fully oval shape. In some species, the seed is winged. The capsules can be harvested once they are dry, but the berries must be fully colored prior to harvest. The seed color is white, grey, and brown to black.

Lions Tail
Agave attenuata

Cabbage Tree
Cordyline australis

Dracaena
Dracaena sp.

Pony Tail
Beaucarnea sp.

Family: AIZOACEAE
Common name: **Carpetweed**
Number of genera: 150
Number of species: 2500
Origin: Tropical and subtropical
Plants: Herbs or low shrubs

The fruit is a capsule that releases a cap or opens lengthwise upon drying. The seed is pea or comma-like in shape, and may be pitted or have bumps. The seed is light, pale to dark brown. Collect the capsules before they open and place in a bag to dry fully. Thresh the seed if required and sieve out the chaff.

Baby Sunrose
Mesembryanthemum sp.

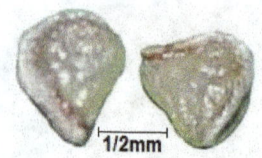

Livingstone Daisy
Dorotheanthus sp.

Family: ALOEACEAE
Common name: **Aloe**
Number of genera: 8
Number of species: 500
Origin: African
Plants: Shrubby perennial evergreens

The fruit is a cylindrical capsule which splits lengthwise as it dries. Each capsule contains numerous seed that are flat or triangular often winged. The seed can be harvested as the fruit opens. The seed is brown to black in color.

Barbados Aloe *Aloe vera*

Family: AMARANTHACEAE
Common name: **Amaranth**
Number of genera: 65
Number of species: 900
Origin: Tropical and subtropical
Plants: Annual herbs

Many of these plants have edible seeds and foliage. The fruit is an aggregate of single-seeded capsules or nuts with variable seed, usually brown or black.

The seed can be harvested once the plants are mature and the flowers have died. Harvest part or all of the plant and place on a ground sheet. Allow a week or so to dry fully before threshing out the seed.

Mukunu-wenna
Alternanthera triandra

Cockscomb
Celosia sp.

Bachelor Button
Gomphrena sp.

Lemon Mint
Monarda citriodora

Family: AMARYLLIDACEAE
Common name: **Amaryllis**
Number of genera: 85
Number of species: 1100
Origin: Tropical and subtropical
Plants: Perennial or biennial herbs

The flowers of these plants are similar in appearance to the lilies and many have bulbous roots. The fruits are drupes or capsules containing one or a few seeds that are corky in appearance. Collect the drupes when they are fully ripe and colored, and remove any flesh. The capsules should be fully matured and dry before collection.

Fruit   Seed

Fireball
Haemanthus multiflorus

Family: APOCYNACEAE
Common name: **Dogbane**
Number of genera: 220
Number of species: 2100
Origin: Tropical
Plants: Trees, shrubs, herbs, and vines

Most of these plants have a white milky sap, and many are poisonous. The fruits are fleshy berries or pod-like capsules that are often long and thin. These can be harvested upon maturity or when they start to split lengthwise. The seed is flattened oval to oblong, sometimes with short or long silky hairs. The seed color ranges from light brown to black.

Oleander *Nerium oleander*

Desert Rose *Adenium* sp.

Frangipani *Plumeria lutea*

Cook Tree *Cascabela thevetia*

Periwinkle *Vinca major*

Family: ARISTOLOCHIACEAE
Common name: **Birthwort**
Number of genera: 10
Number of species: 600
Origin: Worldwide
Plants: Vines with some shrubs and herbs

The fruits are capsules that open into 4 to 6 equal valves. The open seed capsule appears similar to an inverted parachute (*Aristolochia*) and the seeds can be collected at this stage. The seeds are numerous and flattened. They are oval to triangular in appearance, and sometimes winged and smooth.

Indian Birthwort
Aristolochia tegarla

Family: ASCLEPIADACEAE
Common name: **Milkweed**
Number of genera: 250
Number of species: 2500
Origin: Subtropical
Plants: Shrubs, herbs, and vines

Many of these plants have a white milky sap. The fruits can be long, cylindrical twin pods to small ovate capsules. The seeds can be narrowly oblong, flattened, or rounded with a tuft of hairs at one end. The seeds are colored brown to black, and can be collected as the pods or capsules change color and start to split.

Wax Flower *Stephanotis* sp.

Family: ASTERACEAE
Common name: **Daisy**
Number of genera: 1100
Number of species: 20,000
Origin: Worldwide
Plants: Annual and perennial herbs with a few trees and shrubs

The daisies are the largest of the plant families. The fruit can be a burr or a group of closely packed singular seeds (achene), or sometimes a drupe. The seeds are slightly or largely elongated with grooves, pits, or wings. In the burrs, the top of the seed has 2, 3, 4 or more spikes. Most of the remaining genera have seed with a tuft of long hairs. The seeds are collected as the burrs or achene's dry or the drupe matures.

Cornflower
Centaurea sp.

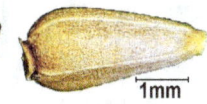

China Aster
Callistephus sp.

Mexican Aster *Cosmos* sp.

Giant Russian Sunflower
Helianthus annuus

Zinnia
Zinnia sp.

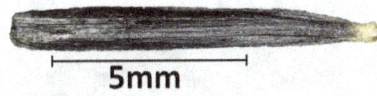

Marigold *Calendula* sp.

African Daisy *Gerbera* sp.

Pussy Foot
Ageratum sp.

Easter Daisy
Aster sp.

Paper Daisy
Helichrysum
sp.

Shasta Daisy
Chrysanthemum
sp.

Dahlia *Dahlia* sp.

English Daisy
Bellis sp.

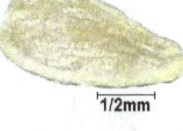

Brachycome
Brachycome sp.

Black-eyed Susan
Gazania sp.

Singapore Daisy
Sphagneticola sp.

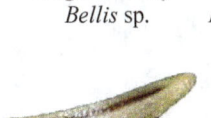

Cineraria
Senecio sp.

Corepsis
Corepsis sp.

11

Blanket Flower
Gaillardia sp.

Common Teasel
Eupatorium sp.

Family: BALSAMINACEAE
Common name: **Balsam**
Number of genera: 4
Number of species: 600
Origin: Tropical to temperate
Plants: Herbs with some shrubs

The fruits can be fleshy or non-fleshy capsules or berries. The seeds can be collected as the fruits ripen and change color.

Impatienes *Impatienes sp.*

Family: BEGONIACEAE
Common name: **Begonia**
Number of genera: 5
Number of species: 1400
Origin: Tropical and subtropical
Plants: Herbs with some climbers and shrubs

The fruit is a capsule; the seeds are minute, often termed "dust seed."

The seeds have some surface sculpturing and the capsules are collected individually. Lightly thresh once fully dry, and sieve out the seed with a fine mesh.

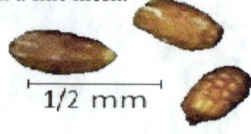

Wax Begonia
Begonia semperflorens

Family: BIGNONIACEAE
Common name: **Bignonia**
Number of genera: 100
Number of species: 800
Origin: Tropical and subtropical
Plants: Trees, shrubs, and vines with some herbs.

The fruits are bean-like capsules which split lengthwise as they ripen and change color from green to brown. They have few to numerous seeds that are often flat, papery, and winged.

Trumpet Creeper
Campsis radicans

Bower Vine
Pandora jasminoides

Family: BORAGINACEAE
Common name: **Borage**
Number of genera: 146
Number of species: 2000
Origin: Worldwide
Plants: Trees, shrubs and herbs

The fruits are capsules, or sometimes drupes or nutlets. Some seeds have barbed hooks that attach to passing animals or objects. The capsules will often open to release their seeds before they are noticed. Collect the capsules as they are encountered and save until a suitable quantity is ready to be sieved of chaff.

1mm

2mm

Echium
Echium wildprettii

Wax Plant
Cerinthe major

1/2mm

5mm

Candytuft
Iberis umbellata

Honesty
Lunaria annua

1/2mm

2mm

Stock
Matthiola sp.

Wallflower
Cheiranthus sp.

Family: BRASSICACEAE
Common name: **Mustard**
Number of genera: 375
Number of species: 3200
Origin: Worldwide
Plants: Herbs with some sub-shrubs

The family is formally called CRU-CIFERAE. The fruits are cylindrical capsules called silique. They dry prior to splitting into two equal halves, with a partition down the middle releasing numerous seeds. The seeds are brown to black in color.

The pods can be collected individually as they ripen, or the fruiting stems can be collected whole. These can be placed in a material or paper bag or on a ground sheet to dry.

Family: CACTACEAE
Common name: **Cactus**
Number of genera: 85
Number of species: 800
Origin: Americas
Plants: Fleshy, spiny trees, shrubs, or herbs

The fruit is a fleshy, colorful berry which can be large and contain few to many seeds. The seeds are mostly round, compressed, smooth, or pitted. They are collected once the fruit is ripe and soft. Collect the ripe berries individually and gently squeeze the pulp into a sieve. With slow running water, lightly press the pulp through the sieve with your fingertip. After the seed is sieved, spread on a non-plastic surface to dry.

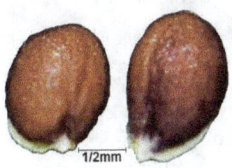

1/2mm

Madword *Alyssum* sp.

1/2mm

1/2mm

Melon Cactus
Melocactus sp.

Torch Thistle
Cereus sp.

13

Family: CAESALPINIACEAE
Common name: **Bean**
Number of genera: 150
Number of species: 2200
Origin: Worldwide
Plants: Trees and shrubs with some herbs and climbers

The beans are formally combined with the family FABACEAE. The fruit is a legume that is generally bean-shaped and can be woody to thin and papery, containing several to numerous seeds.

    The seed is often kidney-shaped, but can also be spherical to flat and oval, as well as round. The color can be reddish, grey, brown, or black, and sometimes mottled. The seeds are collected from the dry pods; the pods often require threshing to release the seeds.

Desert Pea
Clianthus formosus

Butterfly Pea
Clitoria literalis

Family: CAMPANULACEAE
Common name: **Bellflower**
Number of genera: 35
Number of species: 600
Origin: Worldwide
Plants: Herbs with a milky sap

The fruits are capsules that open progressively, with segments in the capsule producing few to numerous seeds. The brown seeds are minute, elliptic, and flattened. Collect the capsules as they begin to dry and split. Thresh if required, and sieve out the seed.

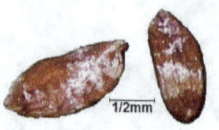 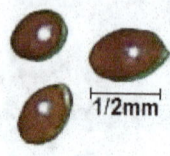

Bellflower
Campanula medium

String of Pearls
Lobelia pendula

Family: CANNACEAE
Common name: **Canna**
Number of genera: 1
Number of species: 50
Origin: Tropical and subtropical
Plants: All are herbs

The fruit is a thin papery capsule, with large, round, hard, dark brown to black seeds.

    Collect the capsules once they are fully dry. Often the capsules retain their seeds for long periods of time if they are undisturbed.

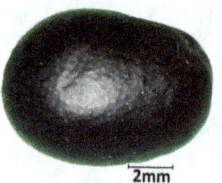

Canna *Canna indica*

Family: CARYOPHYLLACEAE
Common name: **Pink**
Number of genera: 75
Number of species: 2000
Origin: Worldwide
Plants: Herbs with some shrubs

The fruit is a very small round capsule measuring 0.5 – 3.5mm with one to numerous seeds that are slightly flattened. The capsules are collected once dry and lightly threshed to release the seed.

Carnation	**Baby's Breath**
Dianthus sp.	*Gypsophila* sp.

Wandering Jew
Commelina sp.

Family: COMBRETACEAE
Common name: **Indian Almond**
Number of genera: 20
Number of species: 500
Origin: Tropics and subtropics
Plants: Trees, shrubs, and vines

The fruit is a winged or ridged drupe containing one seed / kernel. The whole fruit is collected individually. No threshing or cleaning is usually required.

Rangoon *Quisqualis* sp.

Family: COMMELINACEAE
Common name: **Spiderwort**
Number of genera: 50
Number of species: 700
Origin: Tropical and subtropical
Plants: Herbs with some climbers

The fruit is a capsule that usually splits lengthwise upon drying. The seeds are small, round, and hard, usually light brown to black. Collect the capsules as they dry and remove the seeds from the capsules individually or by threshing.

Family: CONVOLVULACEAE
Common name: **Morning Glory**
Number of genera: 50
Number of species: 1500
Origin: Tropical and subtropical
Plants: Herbaceous climbers with some herbs and shrubs

The fruit is a capsule that splits lengthwise upon drying. It is also sometimes a berry or nut. The seeds can be round, triangular, or oval, smooth, and sometimes hairy. They are brown to black in color.

The capsules can be collected from the plants individually as they split and dry. Threshing is often required to release the seeds.

The chaff of some species can cause extreme skin irritation, so protective gloves and a face mask are usually required.

Morning Glories

Merremia sp.   *Convolvulus* sp.

Exotic Love *Mina lobata*

15

Family: CRASSULACEAE
Common name: **Stonecrop**
Number of genera: 25
Number of species: 900
Origin: Worldwide
Plants: Shrubs and herbs with some trees

The fruit is an ovate to cylindrical capsule, producing two to numerous small brown to black seeds. *Bryophyllum delagoense* (mother-of-millions) is a significant invasive weed species. Collect the capsules as they dry and split and thresh if required before sieving out the seeds from the chaff.

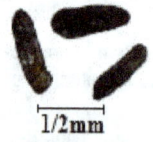

Chandelier Plant
Bryophyllum sp.

Family: DIPSACACEAE
Common name: **Teasel**
Number of genera: 11
Number of species: 350
Origin: Temperate
Plants: Shrubs and herbs

The fruits are dry, single-seeded drupes. The whole fruits are collected individually once dry and sieved of any clinging materials.

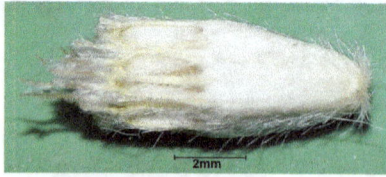

Pin Cushion Flower
Scabiosa sp.

Family: DROSERACEAE
Common name: **Sundew**
Number of genera: 4
Number of species: 100
Origin: Worldwide
Plants: All are herbs

The fruits are capsules that split upon drying, releasing minute seed that can be round or linear and are sometimes winged. Collect entire capsules as they change color and begin to split. Sieve out the loose seed and remove the empty capsules.

Sundew *Drosera* sp.

Family: EUPHORBIACEAE
Common name: **Spurge**
Number of genera: 300
Number of species: 7500
Origin: Worldwide
Plants: Shrubs and herbs

The spurges have a white milky sap. The fruit is usually a dry capsule that splits lengthwise upon drying; it is sometimes a drupe or berry.

The seeds can be oval, triangular, or round and can be smooth or pitted. The drupes or berries are collected once ripe and the capsules are collected as they split.

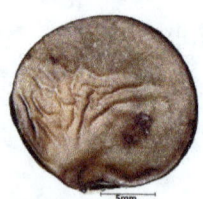

Petty Spurge *Euphorbia* sp.

Family: GENTIANACEAE
Common name: **Gentian**
Number of genera: 80
Number of species: 1100
Origin: Temperate
Plants: Small trees, shrubs, herbs and vines

The fruit is a small capsule, often with minute, numerous oval-pitted seeds which vary from yellow to reddish-brown to black in color. Collect the capsules as they dry and begin to split. Thresh if required and sieve the chaff from the seeds.

Prairie Gentia *Eustoma* sp.

Family: GERANIACEAE
Common name: **Geranium**
Number of genera: 5
Number of species: 750
Origin: Worldwide
Plants: All are herbs

The geraniums are known by their distinctive bird's bill seed pod. The seeds can be oval, cylindrical with a corkscrew-like appendage, or cylindrical with fine hairs. Collect the whole pods individually as they dry and thresh to release the seeds.

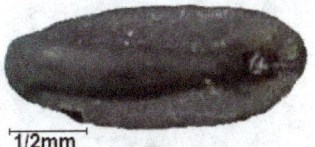

Austral Stork's Bill
Pelargonium australi

Family: GESNERIACEAE
Common name: **African Violet**
Number of genera: 150
Number of species: 3200
Origin: Tropical and subtropical
Plants: Small shrubs, perennial herbs, and vines

The fruits are berries or cylindrical capsules with small to minute seeds.

The berries are collected individually as they ripen and can be pressed gently through a sieve with slow running water. The capsules are collected as they begin to dry and are threshed to release their seeds.

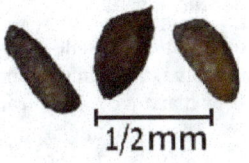

Nodding Violet
Streptocarpus caulescens

Family: GOODENIACEAE
Common name: **Goodenea**
Number of genera: 17
Number of species: 400
Origin: Southern Hemisphere
Plants: Shrubs, herbs, and some climbers

The fruit is often a cylindrical capsule. Some are soft and fleshy drupes, while others are hard nuts.

The seeds are variable, some being flat with a wing while others are oval, elongated, or angular. The color of the seeds is variable, ranging from yellow or brown to black and shiny. Collect the fruits once fully ripe and treat depending on the type of fruit encountered.

17

Blue Pincushion
Brunonia australis

Family: HAEMODORACEAE
Common name: **Bloodwort**
Number of genera: 14
Number of species: 100
Origin: Southern Hemisphere
Plants: Perennial herbs

The fruit is a small capsule that splits upon maturity, containing numerous small seeds that are ovoid to teardrop in shape. They are often pitted or furrowed and are red, brown, grey, or black in color. Collect the capsules as they begin to dry and thresh if required. Sieve out any chaff before storing the seeds

Kangaroo Paw
Anigozanthos sp.

Family: HYDROPHYLLACEAE
Common name: **Waterleaf**
Number of genera: 20
Number of species: 300
Origin: Tropical to temperate
Plants: Herbs with some spiny shrubs

The fruits are capsules that split open upon drying. Collect the capsules and thresh lightly if required. The seeds can be sieved of any chaff before storing.

Baby Blue Eyes	Fluffy Lavender
Nemophila insignis	*Phacelia tanacetifolia*

Family: IRIDACEAE
Common name: **Iris**
Number of genera: 85
Number of species: 1500
Origin: Tropical to temperate
Plants: Perennial herbs with some small shrubs

The seed capsule is a cylindrical tube about 2cm in length; this length will vary with the various genera and species being collected.

As the capsule dries, it splits lengthwise into three equal parts. The seeds are mature at this stage and can be harvested.

The seeds are generally oval and flat or angular, sometimes with a distinct beak, sometimes pitted, smooth, or rough. They are brown to black in color.

Iris	Freesia
Dietes sp.	*Freesia* sp.

Leafy Purple Glag
Patersonia glabrata

Family: LAMIACEAE, LABIATAE
Common name: **Mint**
Number of genera: 180
Number of species: 3500
Origin: Worldwide
Plants: Aromatic herbs with some trees and shrubs

The fruit is usually a dry capsule with four very small seeds. The seeds from the different genera are highly variable. Collect the capsules as they dry and sieve out the seeds. Alternatively, place a ground sheet under a plant and check for seeds daily.

1mm

Lechenaultia
Lechenaultia sp.

1/2 mm

Coleus
Coleus sp.

1mm

Cats Whiskers
Orthosiphon aristata

1/2mm

Catmint
Nepeta cataria

1/2mm

Westringia
Westringia sp.

1mm

Bells of Ireland
Moluccella sp.

Family: LARDIZABALACEAE
Common name: **Lardizabala**
Number of genera: 8
Number of species: 35
Origin: Subtropical to temperate
Plants: Climbers with some shrubs

The fruits can be a tight cluster of fleshy fruits, known as an aggregate of berries. Collect the entire aggregate of berries and gently remove the flesh and pulp. This can be undertaken using a large sieve. Gently press out the flesh and pulp, leaving the seed. Slow running water is helpful to keep the pulp moving through the sieve.

1mm

Five Leaf *Akebia gumata*

Family: LILIACEAE
Common name: **Lily**
Number of genera: 280
Number of species: 4000
Origin: Worldwide
Plants: Perennial herbs with some evergreen shrubs and climbers

The fruit can be a capsule or a colorful berry. The seeds of this family are generally smooth, with some being winged and others having a variety of ornamentation. The seeds are shiny orange to red or grey, brown, or black, and are approximately 2 to 6mm.

Collect the berry or capsule once fully ripe. The seeds from the berries can be cleaned by gently pressing the pulp through a sieve with your fingertip and slow running water. The capsules can be lightly threshed to remove the seed.

Lily
Lilium sp.

Flax Lily
Dianella sp.

Daylily   *Hemerocallia* sp.

Family: LIMNANTHACEAE
Common name: **Meadowfoam**
Number of genera: 2
Number of species: 11
Origin: Temperate
Plants: All are herbs

The small fruits are an aggregate of nutlets. Collect the aggregate of fruit and thresh to free the nutlets from the stems. Sieve out the chaff and store the nutlets once fully dry.

Poached Egg Plant
Limnanthes douglasii

Family: LOGANIACEAE
Common name: **Logania**
Number of genera: 29
Number of species: 500
Origin: Tropical and subtropical
Plants: Tree, shrubs, herbs and woody climbers

The small fruits are berries, drupes, or capsules containing one to many seed. The seed is small to minute and often pitted, lumpy, or smooth. The seeds are brown to black and can be collected from the small fruits once they are fully ripe or from the capsules as they change color and begin to dry.

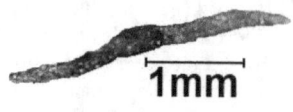

Butterfly Bush
Buddleia davidii

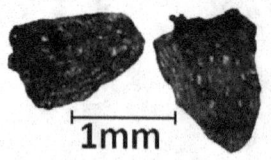

Cabbage Tree
Fagraea crenulata

Family: MALVACEAE
Common name: **Mallow**
Number of genera: 75
Number of species: 1500
Origin: Worldwide
Plants: Perennial herbs and shrubs

The fruit is a capsule which splits from the top, or sometimes a berry. The seeds are somewhat kidney-shaped, and are occasionally winged or with horns. They are brown or black in color.

The capsules can be collected once they have dried. They are then threshed to release their seed. Sieve out the chaff and store the seed once dry.

20

Hairy Hollyhock
Alcea rugosa

Family: MELASTOMATACEAE
Common name: **Melastoma**
Number of genera: 250
Number of species: 4500
Origin: Tropical and subtropical
Plants: Shrubs and herbs

The fruit is a capsule or a berry. The seeds are small and numerous; they are often oval and pitted. Collect the berry or capsule once fully ripe. The seeds from the berries can be cleaned by gently pressing the pulp through a sieve with your fingertip and slow running water. The capsules can be lightly threshed to remove the seed.

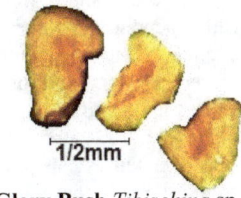

Glory Bush *Tibischina* sp.

Family: MENYANTHACEAE
Common name: **Buckbean**
Number of genera: 5
Number of species: 70
Origin: Worldwide
Plants: Aquatic herbs

These plants usually grow where the land is inundated with rain annually. The fruit is a small capsule that can be collected after the water has subsided and the land begins to dry.

Collect the entire plant or the capsules individually, and remove any dirt before lightly threshing out the seed.

Australian Marshwort
Nymphoides crenata

Family: MYRTACEAE
Common name: **Myrtle**
Number of genera: 155
Number of species: 3500
Origin: Tropical regions and Australasia
Plants: Trees and shrubs

All have oil glands in their leaves and are predominantly evergreens. The fruits are capsules that can be papery or hard. They split at the front of the capsule upon maturity, releasing the seed; some capsules may remain unopened for several years before releasing the seed. The seed can be saucer-shaped, with a wing all around, or wingless; they can have a wing on one end, be a pyramid shape, or be crescent-shaped or straight. Color ranges from reddish-brown to brown or black. Collect the capsules once splits appear in the front or the outer rim colors. Spread the capsules on a ground sheet and allow them to dry fully before sieving.

Scarlet Kunzea
Kunzea baxterii

Beaufortia
Beaufortia purpurea

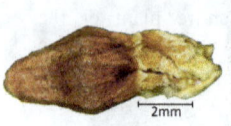

**Geralton Wax
Flower**
*Chamelaucium
uncinatum*

**Swan River
Myrtle**
*Hypocalymma
robustrum*

Family: NYCTAGINACEAE
Common name: **Four O'clock**
Number of genera: 33
Number of species: 290
Origin: Tropical and subtropical
Plants: Trees, shrubs, herbs or
climbers

The dry, non-fleshy drupes have one
seed. Collect the drupes individually
once they are fully ripe. Threshing
should only be required once the
fruits have dried to release anything
attached to the seed. Clean and sieve
out any chaff.

Four O'clock *Mirabilis jalapa*

Family: NYMPHAEACEAE
Common name: **Water Lilies**
Number of genera: 5
Number of species: 50
Origin: Tropical and subtropical
Plants: All are aquatic herbs

The fruits are spongy berries containing one to many round or elliptical
seeds, which are sometimes hairy.
Collect the ripe berries from the old
flower stems. These are located just

below the water line, often under the
leaves. Allow a little time for the
berries to dry before removing the
seeds by hand. Use a bowl of water to
aid in cleaning the seeds before drying them.

Lotus Lily
Nelumbo sp.

Water Lily
Nymphaea sp.

Family: ONAGRACEAE
Common name: **Evening Primrose**
Number of genera: 20
Number of species: 650
Origin: Worldwide
Plants: Tree, shrubs, and herbs

The fruits are usually capsules that
may or may not open as they dry, but
they can also be berries or nuts. The
quantity of seeds ranges from 2 to
over 100 depending on which fruit is
encountered. Some seeds are non-
hairy, others are hairy, and some have
a tuft of bristles at one end.

Collect the fruits once they are
fully ripe. The berries can be cleaned
by pressing the pulp through a sieve.
The nuts should not require any
threshing. Capsules can be threshed
to release their seeds before sieving
off the chaff.

Godetia *Clarkia amoena*

Family: ORCHIDACEAE
Common name: **Orchid**
Number of genera: 730
Number of species: 20,000
Origin: Worldwide
Plants: Epiphytic or terrestrial peren-
nial herbs with some climbers

The orchids are one of the largest
plant families. The fruit is an elliptic
capsule that splits in several places
lengthwise as it ripens, releasing
numerous seeds that can be minute or
up to 5mm in length. They are white
to light brown in color.

Collect the whole capsule as it
splits. The seed can be sieved easily
from the capsule with no threshing
required.

Crucifix Orchid	Vanda Orchid
Epidendrum sp.	*Vands* sp.

Family: OXALIDACEAE
Common name: **Sorrel**
Number of genera: 8
Number of species: 900
Origin: Worldwide
Plants: Herbs with some trees and
shrubs

The fruit is a capsule that comprises
of five sections, each containing several seeds. The capsules can open
explosively when the right environmental conditions are present.

Eye injuries are not uncommon,
as the capsules can explode in your
hands. Use protective glasses when
collecting seed from this family.

Three Leaf Clover *Oxalis* sp.

Family: PAPAVERACEAE
Common name: **Poppy**
Number of genera: 44
Number of species: 760
Origin: Worldwide
Plants: Herbs with a few trees and
shrubs

The fruit is a capsule containing numerous seeds which are small and
round. They are often brown to black
in color.

Collect the whole capsule and allow time for them to dry before
threshing and cleaning the seed.

Californian Poppy	Tulip Poppy
Eschscholzia sp.	*Hunnemannia fumariifolia*

Family: PAPILIONACEAE
Common name: **Legume**, **Pea,** or
Pulse
Number of genera: 400
Number of species: 12000
Origin: Worldwide
Plants: Trees, shrubs, herbs, and
vines

23

The family is formally referred to as FABOIDEAE, FABACEAE, or LEGUMINOSAE. The fruit is a legume with two equal halves that opens upon drying, sometimes dispersing the seed with intense force.

The seed size and shape is variable, generally cylindrical or a flattened oval, and sometimes with hairs. Collect the whole legume and allow time to dry fully. The legumes may require covering to contain the seed from explosive species.

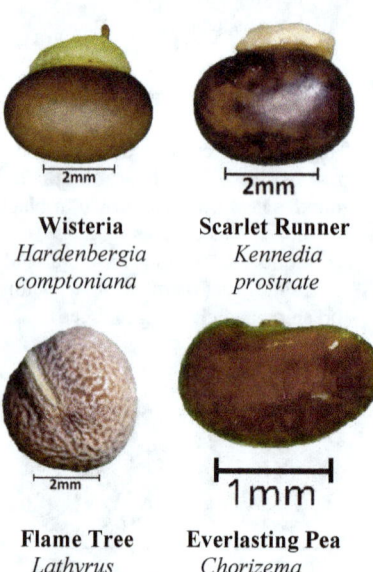

Wisteria
Hardenbergia comptoniana

Scarlet Runner
Kennedia prostrate

Flame Tree
Lathyrus latifolius

Everlasting Pea
Chorizema cordatum

Family: PASSIFLORACEAE
Common name: **Passionfruit**
Number of genera: 10
Number of species: 500
Origin: Tropical to subtropical
Plants: All are vines

The fruits are thin-walled capsules or berries. The seeds are spearhead shaped, pitted or scalloped, and are usually surrounded by a fleshy pulp. The fruit is collected individually and the pulp removed.

The seed is brown to black in color and can be cleaned by pressing the pulp lightly through a sieve with your fingertip and slow running water.

Passion Flower
Passiflora coccinea

Family: PITTOSPORACEAE
Common name: **Pittosporum**
Number of genera: 9
Number of species: 200
Origin: Tropical and subtropical regions of the southern Hemisphere.
Plants: Evergreen trees and shrubs

The fruit is a capsule or berry that splits from bottom to top, revealing a brown or black seed which is often surrounded by a colorful pulp. The seed is variable in size and shape and can be cleaned from the pulp if required. The thin pulp is left to dry on many of these species.

Native Frangipani
Hymenosporum flavum

Family: PLUMBAGINACEAE
Common name: **Leadwort**
Number of genera: 24
Number of species: 560
Origin: Worldwide
Plants: Herbs, shrubs and semi climbers

The fruit is a sticky capsule with soft spines that will adhere to anything that passes; each capsule contains one seed. The seed can be removed from the capsule manually or the capsules can be left to dry fully before being threshed.

2mm

Leadwort
Plumbago auriculata

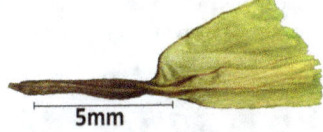

5mm

Statice *Limonium latifolium*

Family: POLEMONIACEAE
Common name: **Phlox**
Number of genera: 25
Number of species: 400
Origin: Northern Hemisphere and South America
Plants: Annual or perennial herbs with some perennial trees and shrubs

The fruit is a capsule that splits into three equal parts releasing one to many seeds. These may be winged. The seed becomes sticky when wet. Collect the capsules individually and thresh before sieving out the seed.

1mm

Cup-and-saucer Vine
Cobea scandens

Phlox
Drummondi cuspidate

Family: PORTULACACEAE
Common name: **Pigweed**
Number of genera: 20
Number of species: 500
Origin: Worldwide
Plants: Annual or perennial herbs

The fruit is a small capsule containing one to numerous smooth or pitted grey or black seeds. The seeds can be collected by harvesting part or all of the plant and placing it on a ground sheet, or by placing the ground sheet under the plants and checking regularly.

1/2mm   **1mm**

Sun Plant
Portulaca sp.

Flameflower
Talinum sp.

Family: PRIMULACEAE
Common name: **Primrose**
Number of genera: 30
Number of species: 1000
Origin: Temperate regions of the Northern Hemisphere
Plants: Annual or perennial herbs

The fruit is a small oval capsule opening at the top containing minute three-sided brown seeds. Collect the capsules once dry and thresh lightly before sieving out the chaff.

25

Lollipops *Primula malacoides*

Family: RANUNCULACEAE
Common name: **Buttercup**
Number of genera: 80
Number of species: 2500
Origin: Worldwide
Plants: All are herbs

The fruit is a berry or a five-sided capsule.

Collect the berries or capsules once fully ripe. The seeds from the berries can be collected by gently squeezing the pulp into a sieve and lightly pressing the pulp through the sieve with your fingertip and slow running water. The capsules can be threshed to release the seed before sieving out the chaff.

Love-in-a-mist
Nigella damascena

 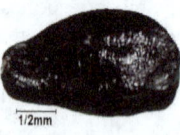

Larkspur   **Granny's Bonnet**
Delphinium sp.   *Aquilegia* sp.

Family: ROSACEAE
Common name: **Rose**
Number of genera: 122
Number of species: 3350
Origin: Worldwide
Plants: Trees and shrubs with some herbs

The fruits of the roses are extremely varied and can be pomes, drupes, hips, achenes, or other named fruit types. The seed is also variable in size and shape, ranging from a single large seed through to minute and dust-like seeds.

It is the actual rose that is of interest here. The fruit is called a hip, which is essentially a dry berry. Collect the hip once it is fully colored and carefully open and remove the seed.

As the flowers cross-pollinate readily there are no guaranties that the seedlings will produce the same flowers as the parent plant.

Rose *Rosa* sp.

Family: RUBIACEAE
Common name: **Madder**
Number of genera: 650
Number of species: 5000
Origin: Tropical to subtropical
Plants: Trees, shrubs, herbs and vines

The fruits are fleshy berries or drupes and occasionally dry capsules. There are one to numerous seeds, often embedded in a succulent mass.

The seed can be highly variable in size, shape, and coloration and should be cleaned depending on the type of fruit that you encounter.

Jungle Geranium *Ixora sp.*

Family: SCROPHULARIACEAE
Common name: **Figwort**
Number of genera: 250
Number of species: 5000
Origin: Worldwide
Plants: Shrubs and herbs

The fruits are small capsules that split from the top or sides, producing few to many minute seeds. The seeds are arrow- or kidney-shaped, rectangular, or oval. All are pitted and are brown to black in color. Collect the capsules as they begin to split, and thresh out the seed.

Pouch Nemesia *Nemesia sp.*

Granny's Bonnet	Snap Dragon
Angelonia sp.	*Antirrhinum* sp.

Foxglove	Toadflax
Digitalis sp.	*Linaria* sp.

Penstemon	Chimp-pansy
Penstemon sp.	*Mimulus* sp.

Family: SOLANACEAE
Common name: **Nightshade**
Number of genera: 90
Number of species: 2800
Origin: Americas
Plants: Small trees, shrubs, herbs, and vines

Many of the genera of this family are important agricultural plants. The colorful, fleshy berries can be harvested upon maturity.

The seeds are flattened, curved, and oval, often with pitting. The color ranges from yellow, orange, grey, brown to black.

Collect the whole fruit once fully ripe. The capsules may or may not open upon maturity. Dry capsules can be threshed, whereas fleshy capsules can be cut open and the seed removed manually.

The pulp of berries can be pressed through a sieve or placed in a container of water and left for about a week to ferment, before being dried and sieved clean.

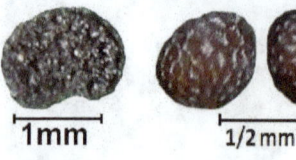

1mm **1/2mm**

Butterfly   **Petunia**
Flower   *Petunia sp.*
Schizanthus sp.

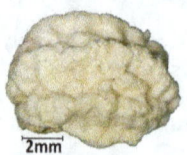

`2mm`

Nasturtium
Tropaeolum majus

Family: THYMELAEACEAE
Common name: **Daphne**
Number of genera: 60
Number of species: 500
Origin: Worldwide
Plants: Small trees or shrubs

The fruit usually remains enclosed at the base of the flower and attached to the flower stem for some time. Occasionally, the fruit is an oval, elliptical, or cylindrical capsule which opens lengthwise as it dries. Seeds can be flattened or pointed, and are sometimes hairy. Collect the whole flower stem and thresh out the seed.

Family: VERBENACEAE
Common name: **Verbena**
Number of genera: 100
Number of species: 3000
Origin: Tropical and subtropical
Plants: Trees, shrubs, herbs, and vines

The oval fruits can be drupes or dry berries that split into halves, thirds, or quarters. Collect the fruit and clean away any flesh and pulp.

Most fruits can be left to dry whole and then lightly threshed to remove any materials attached to the seed.

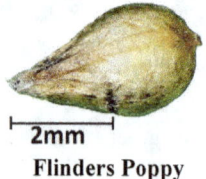

2mm

Flinders Poppy
Pimelea decora

Family: TROPAEOLACEAE
Common name: **Nasturtium**
Number of genera: 3
Number of species: 92
Origin: Tropical to temperate
Plants: All are herbs

The fruits are sometimes fleshy and are like nuts or drupes in appearance. However, these break into 1, 2, or 3 segments upon maturity. The fruit is generally collected individually from the plants and the seed cleaned if required.

1mm   **2mm**

Pigeon Berry   **Spanish Flag**
Duranta   *Lantana*
repens   *camara*

`5mm`

`2mm`

Bleeding Heart   **Woolly**
Glory Bower   **Congea**
Clerodendrum   *Congea*
thomsoniae   *tomentosa*

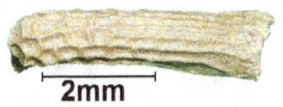

2mm

Vervain *Verbena* sp.

Family: VIOLACEAE
Common name: **Violet**
Number of genera: 22
Number of species: 900
Origin: Worldwide
Plants: Small shrubs and herbs

The fruit is a round to cylindrical capsule or berry containing numerous small seeds. The seeds can be round, flattened, or boat-shaped, and can be pitted or smooth. They are yellow, golden, or brown in color.

Collect the berries once they are fully ripe and gently squeeze the pulp into a sieve. Lightly press the pulp through the sieve with your fingertip and slow running water. Dry the seed before sieving out any remaining chaff.

The capsules can be collected once dry and threshed to release the seed.

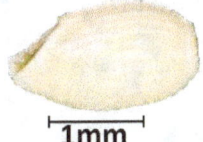

1mm

Shrub Violet
Hybanthus aurantiacus

Family: XANTHORRHOEACEAE
Common name: **Grass Trees**
Number of genera: 10
Number of species: 100
Origin: Endemic to Australia with the exception of two species in New Guinea and one in New Caledonia
Plants: All are perennial grass-like plants

The flower spikes produce numerous fruit capsules which split along their lengths. The seeds can be round, oval, or flattened. Seed color ranges from orange, brown, and red to black.

Collect the entire flower stem or the individual capsules and allow time to dry fully. Some species require threshing to remove the seed from the capsules. Sieve the seed to remove any chaff before storing.

5mm

Grass Tree *Xanthorrhoea* sp.

2mm

Slender Matrush *Lomandra hystrix*

Family: ZINGIBERACEAE
Common name: **Ginger**
Number of genera: 40
Number of species: 1000
Origin: Tropical and subtropical
Plants: All are herbs

The fruit is a capsule or colorful berry. In some species of *Alpinia,* the seed could best be described as "segments of a sphere." The seed is colored brown to black.

Collect the whole fruit once ripe and clean depending on the type of fruit encountered.

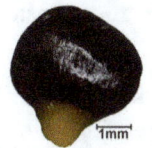

1mm

2mm

Gingers *Alpinia* spp.

Chapter 3
How to Thresh and Clean Seed

Threshing

First, let's look at the term **threshing**.

Threshing is the term used for the breaking up and separation of the dried seed pods, seed heads, or capsules from the seed. The waste from the seed is called **chaff**. It is after threshing that the actual cleaning of seed occurs. These are two separate activities. Threshing of one form or another is usually required prior to cleaning.

With the use of some common household items, the threshing of seed is not very difficult. Sieves, colanders, tea strainers, buckets, and bowls can all be employed to get the job done—even by professional collectors. Additional equipment for enthusiastic collectors includes a mechanical thresher or a garden blower-vac.

Threshing can be undertaken by a number of means, including manual and mechanical methods. Both are effective, although an appropriate method should be sought depending on the amount of material that re-quires threshing. There is little point in using a mechanical thresher for a handful of material; conversely, you really don't want to thresh a truck-load of material by hand!

Hand Threshing

As implied by the name, this is using your hands to crush the chaff from around the seed. This can be undertaken by placing the material in a bucket or bowl, and crunching it up until the seed is freed from the capsules. Gloves are generally required for this method, as the action required can hurt your hands.

Unfortunately, this is not always an effective technique for many of the harder woody species, as the pods are simply too difficult to crush manually.

After you have finished hand threshing the seed / chaff material, gently but firmly swirl the contents around in a bucket or bowl. The usually heavier seed will work its way to the bottom, allowing the bulk of the chaff to be removed and discarded. This makes the cleaning process easier. Should the seed and chaff not separate; cleaning will be more complicated and time-consuming.

Threshing with Sieves

Threshing can be undertaken by simply pressing the collected material through an appropriate sieve. This works well for many species with relatively brittle capsules or pods.

To determine which sieve is best suited for the material being threshed, select a sieve with holes slightly larger than the seed.

Good quality gloves are recommended when using this method, as injuries can occur from small sticks, thorns, etc. piercing the skin.

Attaching a sieve to a child's swing can aid greatly in this method of threshing, as it enables you to use a much larger sieve.

Threshing with a Towel or Rag

This method is restricted for the seed of fruit and berry species where the outer layer of the seed peels away like a skin. The cucurbits are one such group where this method is well-suited.

To use this method, rub the seeds between two layers of material, such as calico or rags. Place only small amounts of seed between the materials at a time, and rub back and forth to remove the chaff.

After you are satisfied the seed and chaff are separated, pour everything into a container to be cleaned.

Mechanical Threshing

There are many mechanical threshers on the market, both manual and motorized. Regardless of type, they are expensive to purchase. You would need to be doing large quantities of threshing to make the purchase of one worthwhile.

Manual, hand-operated thresher

Cleaning Seed

The cleaning of seed after it has been threshed can be a very rewarding experience, as you get to see the end product of your labors. Cleaning, at its most basic, needs only to involve the use of two bowls and the wind; at the more complicated end of the spectrum, it may employ specialized equipment, such as sieves or a motorized cleaner.

Regardless of what equipment is required, cleaning the seed helps reduce the space required for storage, reduce or eliminate pest problems, and makes replanting much easier.

Two Bowl Method

The most simple and inexpensive way to clean seed is the aptly named Two Bowl Method. First, find two suitable bowls; these can be medium-sized plastic food containers, 10-liter

31

water buckets, etc. Use your imagination!

Slowly tip the uncleaned seed (seed and chaff mixture) from the first bowl into the bowl underneath, allowing the wind to blow the chaff away. Wind can be provided either naturally or via a fan. The distance between the bowls will require adjustment, depending on wind strength. Swap bowls and continue tipping the seed from one bowl to the other until you are satisfied with the cleanliness of the seed. In some cases, the seed will not become 100% clean, and a final clean using tweezers is required. Amaranth and eucalyptus are good examples of seeds that are left only partly clean using this method.

Sieves and Sieve Sizes

Having several sieve sizes is always helpful, although purchasing an analytical quality sieve (as pictured) can be prohibitively expensive.

Flour sieves

Assorted kitchen sieves

Analytical sieve

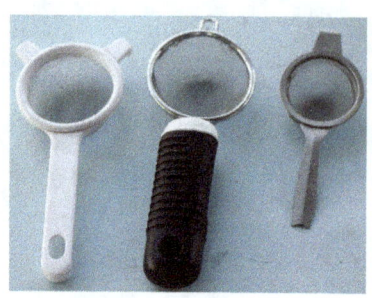

Assorted tea strainers

There are, however, alternatives to using expensive analytical sieves. The most cost-effective sieves can be found in almost every kitchen. These are basic flour and kitchen sieves, tea strainers, and colanders, all of which are available at your local store.

Other Sieves

There are many household items that can be used effectively as sieves for cleaning seed. Mosquito meshing comes in a handy size, whilst a colander is excellent for some of the larger seeds. Audio speakers, both car and home, provide mesh that makes an excellent sieve.

The material used for the mesh of a sieve is limited only by your imagination.

Speaker cover

Cleaning Pods and Dry Capsules

After capsules or pods have been threshed, the cleaning is greatly expedited using sieves. Often two or more sieves are used to separate the larger and smaller portions of the chaff from the seed.

Use several wide-mouthed bowls whilst sieving the seed, and have a bucket handy into which to place the waste. Doing this helps in chaff separation, and also contains any spillage in the event of an accident, making re-sieving easier.

Start with a sieve that has a mesh size several sizes larger than the seed. The reasoning behind this is to remove the larger unwanted materials as efficiently as possible; this also reduces the volume of materials remaining to be cleaned.

After the bulk has been removed, use a sieve that is only slightly larger than the seed and slowly sieve the seed through. There will usually be several seeds remaining that are bigger than usual; collect these to add back to the cleaned seed once finished. Repeat this process until no large chaff can be effectively removed.

Now use a sieve slightly smaller than the seed and sieve out the smaller material. Often, small seed will be sieved out. Reject these, as they are usually immature or inferior. It is more beneficial to focus on the larger, healthy seed.

Repeat this process until the seed is as clean as possible. The final clean can be undertaken utilizing the wind or a mechanical cleaner.

Cleaning Berries and Drupes

The cleaning of fleshy fruits such as berries and drupes are sometimes more interesting than it sounds, as many seed-bearing fruits have their own unique methods of seed extraction.

33

Berries

Berries are fruits with more than one seed, such as fruit from cacti and duranta. Some berries only have a few seeds, while others will contain hundreds. Regardless of how many seeds a berry has, they can usually be cleaned using a sieve and water.

If the size of the seed is unknown, you will need to gently cut open the berry and check the seed. To determine what size sieve will be required, select one that is slightly smaller than the seed. This will allow the pulp of the berry to be washed through the mesh. You may need to press the pulp through the sieve with your fingers or a small spatula.

Once the sieve is selected, place the entire berry or sections of it on the sieve. This process will depend on the size of the berry and sieve. Do not overload the sieve, as it only makes the task more difficult, and may result in the loss of seed, the failure of the sieve, or both.

With the water running slowly over the berry, gently press the pulp through the sieve. The seed should remain in the sieve. If the seed also passes through the sieve, select a smaller mesh size and start again.

Once the seed is considered clean, and the pulp is removed, place the seed on a clean, non-plastic surface and allow it to dry. This process can be expedited by dabbing the seed with a dry cloth, which can be a little tricky if the seed is small, as it tends to stick to the cloth and become a problem.

Spread the seed thinly and avoid mounds, as these are more likely to become moldy. As the seed dries, stir or mix at least daily to aid in thorough drying. This method works well for both small and large seeded species.

Drupes

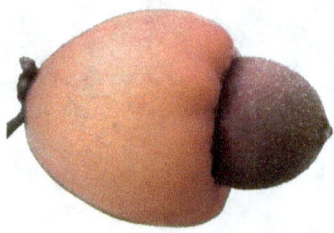

Drupes are single-seeded fruits. The fleshy drupes can be cleaned with a sieve in the same manner as berries. However, cleaning the drupes with flesh that has dried can sometimes be much more difficult.

The very small drupes with dry flesh can be cleaned whole. The entire fruit is left to dry out thoroughly prior to threshing the skins off and then cleaning with a cleaner or sieves.

CHAPTER 4
THE STORAGE
OF SEED

Introduction

Storing seed correctly is the most important step when considering how to collect, clean, and store seed. If seeds are stored correctly, they can last for years. However, if stored poorly, they will lose their viability and fail to germinate once planted.

An important point to remember here is that there are a number of seed-producing species that will only remain viable for a week, sometimes less. These seeds are mostly from fleshy tropical fruits, where there is plenty of moisture and germination occurs quickly. There are also many other plants in arid zones that produce dry capsules with seed that won't store for long. For these plants, storage is not an option, so the collection and cleaning of their seed should be undertaken with the intent of immediate propagation only.

There has been a great deal of research conducted over the last two centuries into the viability of seed under storage conditions, and many good research papers have been published on the findings. This research has shown that many of the seed species that are collected and cleaned correctly can be stored for use at a later date. This is good news for gardening enthusiasts, as it means that most of the plants they are collecting seed from can be preserved.

Regardless of the challenges involved in the storage of seed, it can be a very rewarding experience. All keen gardening enthusiasts should attempt to store the seed from their respective interests to plant in later seasons or exchange with other like-minded enthusiasts.

Take notes on when the seed was stored, how it was stored (in a box, in the refrigerator, etc.), and most importantly, when it was planted and if it germinated. This information on the viability of plant species in your district is very important to plant researchers, as it contributes to the general knowledge on plants worldwide.

Putting Seed into Storage

Only store the best quality seed, i.e., seed that is undamaged, free of defects, pests, chaff, and debris. There are occasions when you cannot collect the best specimens, as the seed simply isn't available due to weather conditions, seasonal factors, or just bad timing. When this occurs, all you can do is the best you can with what you have. Sometimes the seed will be suitable to propagate next season, and sometimes it will fail. This is all part of the enjoyment of collecting seeds.

Seeds should be stored once they are cleaned of all chaff, dirt, and fleshy dried pieces. This helps reduce the volume being stored and reduces the amount of contaminating materials in the seed. Contamination includes both pests and loose materials. Pests include insects, molds, and fungi, all of which can easily destroy stored seed.

Seed should not be put into storage on days when the humidity is high, as the seed may draw moisture from the atmosphere, which can lead to spoilage. This most commonly occurs when seed is stored in plastic bags.

Ideally, seed should be stored once fully dry. This sounds simple enough, but looks can be deceptive, and many seeds have been ruined due to eagerness. Most seeds take about two weeks to fully dry if maintained at an even, warm temperature. They may look dry on the outside, but are often still moist on the inside, so be patient. Many seeds reduce in size as they dry, so also look for this as it occurs.

Once the seeds are ready to be stored, take the time to examine them carefully for any problems, and remove anything that looks suspicious. "When in doubt, throw it out" is a simple expression that is well worth remembering.

Overcoming Seed Pest Problems

The storage of seed has always been an issue due to the many and varied circumstances that can arise without warning. Issues such as humidity, hot summers, cold nights, insects, molds and fungi, and many others add to the difficulty in storing seed successfully.

Seed for storage must be dry, free of dirt, and most importantly, free of insects and their eggs. The weevil eggs pictured below have hatched and the larvae have eaten this bean seed.

Weevil eggs

Weevils and Other Beetles

An excellent way of removing insect eggs from seed, especially large hard seed such as beans, is to make up a dilute solution of bleach (10ml/L)

and soak the seed for ½ an hour. Remove and rinse the seed before thoroughly drying.

Fungicides and insecticides are readily available and can be used to treat seed where problems are too intense to overcome without their use. Take every possible precaution when using these products, as they are **dangerous**. Read the instructions on the label before using any product.

Weevils are one of several insect pests that eat seed. The photo above shows the damage that can happen if seed is stored poorly. A mature weevil is shown below.

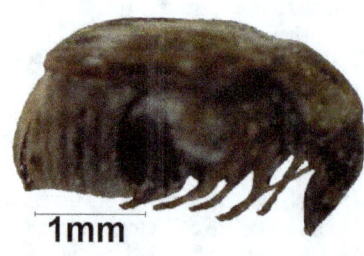

1mm

Weevils and other seed-boring beetles can range in size from almost microscopic to large, impressive specimens. Wood- and seed-boring beetles range in color from drab brown to highly colorful, and many are highly prized by collectors.

Moths

Moth larvae are a major economic pest in cereal crops throughout the world. A great deal of money is spent each year in trying to control the damage and losses caused by these small, troublesome insects.

The specimen shown below is about to pupate after consuming the bulk of a Bauhinia seed. Also shown is an immature larva that was inside the nearby seed when the pod was opened.

Being observant when cleaning your seed will prevent many of the problems that can arise with seed storage. However, it is always a good practice to check your seed regularly and remove anything that may look suspicious.

Fungi

Fungi are the group of plants that lack chlorophyll and leaves. They include molds, mushrooms, mildew, rusts, and smuts.

Of the pest problems encountered, problems with fungi are often the easiest to control and manage.

The problem with molds usually occurs if the seed is put into storage while it is still moist, or if the seed is stored under poor environmental conditions. Whatever the case, if the situation is discovered soon enough, the problem may be reversed, as it takes some time for molds and fungi to penetrate the seed surface and cause damage.

If seed is found with mold growth, you should first remove all affected seed lots from the storage container, as you do not want to infect the remainder of your collection. The affected seed can either be disposed of or cleaned. If the seed is not too badly contaminated, the problem can be solved without disposal.

To clean mold from seed, gently rub the seed with a cloth moistened with bleach solution (10ml/L dilution) until the mold is removed. Staining often remains, but this should not affect the seeds' ability to germinate. This problem will have affected the seeds' long-term storage viability, so replacement should be considered as soon as possible.

Finally, fungicidal powders can be purchased that will inhibit fungal growth. These should be considered if this problem is likely to occur in your location. Remember to read the label and always follow the manufacturer's instructions—these products are **dangerous** when used incorrectly.

Temperature

Few people have access to expensive storage equipment that can maintain a stable temperature at the desired temperature setting. There are two issues here: the temperature at which the seed is stored, and the stability of this temperature. Seeds can be maintained at a low temperature of 2-4^{0}C for long periods, or they can be frozen for storage over centuries. It is the stability of the temperature that is of the greatest importance to the average collector.

As there is no way of maintaining a constant temperature without equipment, we will look at how to maintain a stable temperature using some commonly found items.

Changing temperatures are not good for seed storage. The constant fluctuations from the day and night cycle will cause the most rapid decline of seed viability. This is one problem that must be overcome if you intend storing seed for any length of time.

To achieve optimal results in saving seed, a suitable position with a stable temperature must first be selected. Avoid locations such as external walls and areas near amenities, air conditioners, fireplaces, or any other temperature-altering devices.

Once a storage location is selected, a suitable container in which to house the various lots of seed must be obtained.

Storage containers

Containers that are suitable include, but are not limited to: foam boxes, cardboard boxes, buckets, fiberglass containers, drums, etc. Some storage containers will require modifications to overcome temperature and humidity changes.

Foam boxes are the best choice, as they are already insulated and are often fully sealed. They can be found in numerous sizes, and usually have tight-fitting lids.

Cardboard boxes are good choices; however, finding thick boxes without gaps at the bottom and with tight-fitting lids is sometimes difficult. Additional cardboard can be glued in any places with gaps and can be used to thicken the base, top, and sides as required. Alternatively, line a cardboard box with polystyrene foam—thereby providing the best of both worlds.

Buckets and drums with tight-fitting lids can be used to store seed as long as the inside is properly insulated with foam or other suitable material. Remember that both the bottom and top require insulation. Otherwise, changes in temperature and humidity will still occur inside the container.

Polystyrene and cardboard are excellent insulation materials. Be extremely careful in choosing other insulating materials, such as roof and wall insulation. These items are not designed to be disturbed and are dangerous if handled incorrectly. If you require insulation materials for your chosen container that may be harmful, you should consult an industry expert.

Humidity

As changes in humidity can cause the loss of seed viability, the container that you have chosen to maintain a stable temperature must also be capable of preventing changes in humidity. This is one reason that the lid must be tight-fitting.

Humidity within any dwelling can change dramatically during the daily cycle, especially in tropical regions during the wet season. Even small changes in humidity can have unwanted results on the viability of your stored seed.

Silica gel is a desiccant, and is excellent for protecting your seed from moisture. Some gels are coated with cobalt chloride or other indicators that reveal when their usefulness has expired. You should check these regularly.

All silica gels can be dried and reused numerous times. To do this, place the gel packs on a tray and place in a cooling oven overnight. Ensure the oven is not too hot, as you do not want to burn the packaging of the gel packs.

Silica gels in one form or another can be purchased from chemists, hardware stores, nurseries, and supermarkets. There are a wide range of products on the market, and listing them all would be almost impossible. Water storage granules from nurseries are ideal.

Silica gel packs can also be collected from tablet bottles and within the packaging of electrical items and other miscellaneous products.

Conclusion

The storage of seed can sometimes be challenging. However, the effort is very rewarding, and many varieties of seed can be stored successfully by anyone willing to take the time and effort.

Following simple suggestions, being patient, and taking care to avoid contamination and pests will all lead to the successful storage of seed.

Remember that not all seeds can be stored, so choose your seeds carefully and avoid those that will present problems. As your experience grows, you can become more adventurous and attempt to store more difficult species, taking notes about the methods used and the outcomes.

Many of the rare species in collections around the world today have only survived because plant enthusiasts have saved them. Your efforts could lead to the saving of species for future generations to appreciate.

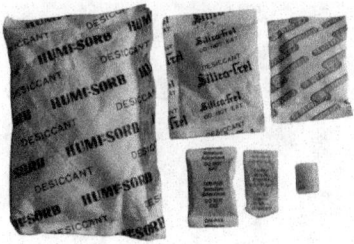

Ensure the gel packs are thoroughly dry before you use them.

REFERENCES

Biles R. E., *The Complete Illustrated Book of Garden Magic.* J. G Ferguson Publishing Company, 1970

Bodkin F., *Encyclopaedia Botanica.* Cornstalk publishing, 1992

Cooper W., *Fruits of the Australian Tropical Rainforest.* Nokomis Edition Pty Ltd, 2004

Delamain B. & Kendall D., *An Introduction to Geraniums in Australia.* Angus & Robertson Publishers, 1987

Fanton M & J., *the Seeds Savers' Handbook.* The seed savers' network, 1993

Fogg H. G. W., *Lilies and their Cultivation.* Gifford, 1961

Gilbert A., *Yates, Green Guide to Gardening.* Harper Collins, 1991

Glass. C, Innes C, Marcus S., *Cacti.* Quintet Publishing Ltd, 1996

Illustrated Guide to Gardening. Readers Digest, 1982

Jones D.L & Gray G., *Climbing Plants in Australia.* Reed Books Pty Ltd, 1988

Macoboy S., *What Flower Is That?.* Lansdowne publishing, 1994

Mann R., *The Ultimate Book of Flowers for Australian Gardens.* Mynah 1995

Novak F. A., *The Pictorial Encyclopedia of Plants and Flowers.* Crown Publishers Inc, 1967

Oakman H., *Tropical and Subtropical Gardening.* Jacaranda Press, 1981

Pilcher M, Davis L, Hurrion D., *Garden Terms.* Hamlyn, 1995

Simpson A. G. W., *Camellias their Colouful Kin and Friends.* Murray book distributors, 1976

Wikipedia, the free encyclopedia

Wrigley J.W & Fagg M., *Banksias, Waratahs & Gravilleas.* Collins Publishers Australia, 1989

Wrigley J.W & Fagg M., *Australian Native Plants.* Angus & Robertson Publishers, 1991

INDEX

41

43

www.ingramcontent.com/pod-product-compliance
Lightning Source LLC
Chambersburg PA
CBHW071143280526
45787CB00003B/1390